DATING DINOSAURS

AND OTHER OLD THINGS

Dating Dinosaurs
AND
Other Old Things

KAREN LIPTAK

THE MILLBROOK PRESS, BROOKFIELD, CT.

Cover photograph courtesy of the Royal Ontario Museum, Toronto, Canada
Illustrations by David Prebenna

Photographs courtesy of: Photo Researchers: pp. 8 (top, Charlie Ott),
15 (left and bottom right, Tom McHugh; top, George Holton; bottom left,
Bucky and Avis Reeves), 45 (top, Stephen J. Krasemann); Peter Arnold, Inc.
pp. 8, 21; Art Resource: pp. 11, 37; Bettmann: pp. 13, 29; U.S. Geological
Survey: p. 18 (top, both); Arizona State Museum: pp. 18 (bottom), 38;
Austin Long, Professor of Geophysics, University of Arizona, Tucson: pp. 28,
35, 53; NASA: p. 30 (both); Religious News Service: p. 35 (right); Larry D.
Agenbroad, Northern Arizona University: p. 40; Laboratory of Tree-Ring
Research, University of Arizona: pp. 45 (bottom), 47 (both); Carolina
Biological Supply: p. 65 (left, both); Richard E. Dodge, Ph.D., Associate
Professor, NOVA University Oceanographic Center: p. 65 (right).

Library of Congress Cataloging-in-Publication Data
Liptak, Karen.
Dating dinosaurs and other old things / by Karen Liptak.
p. cm.
Includes bibliographical references and index.
Summary: Describes various techniques used by scientists to date
objects from the recent and distant past, including dinosaur bones,
rocks, and famous works of art.
ISBN 1-56294-134-8
1. Radioactive dating—Juvenile literature. 2. Archaeological dating—Juvenile literature. [1. Radioactive dating. 2. Archaeological dating.] I. Title.
QE508.L56 1992
930.1'0285—dc20   91-23072 CIP AC

CONTENTS

INTRODUCTION: "HOW OLD IS IT?" 7

THE ABCs OF THE DATING GAME 10

EARLY DATING OF DINOSAURS 14

RADIOACTIVITY AND MODERN DATING 24

COSMIC TIME MARKER: RADIOCARBON DATING 32

SECRETS TREES TELL 42

SOME DATES ARE MAGNETIC 49

NEW WAYS SCIENTISTS GET DATES 55

DATING IN THE FUTURE 64

A DATING VOCABULARY 67

FURTHER READING 70

INDEX 71

INTRODUCTION
"HOW OLD IS IT?"

Whenever a scientist finds the remains of an ancient plant or animal, or an object made by people of long ago, one question always comes up: "How old is it?" A dinosaur's huge leg bone uncovered in England, an early human's jawbone dug up in Ethiopia, a prehistoric American Indian water jug found in the United States—all must be dated.

Scientists must also assign dates to ancient natural events, such as ice ages, earthquakes, and volcanic eruptions. We can't tell where any object or event fits into the ongoing story of life on earth until we know its age.

To help us on our journey into the past, scientists have developed a variety of scientific dating techniques. Some make use of "natural clocks," such as the growth rings of trees. One major dating process relies upon the rate at which some elements break down, or decay. Another method uses clues left behind by changes in the earth's magnetic field. Each of these methods, often in combination with others, has helped clear up mysteries from the past.

Dinosaur bones found in Dinosaur National Monument in Utah.

Natural events, such as volcanic eruptions, produce certain rocks that we can date, in order to establish when such events took place.

When did the first dinosaurs live and the last dinosaurs die? What is the age of the oldest rocks on earth? How old are the rocks astronauts brought back from the moon? When did the last ice age begin? These are some of the questions scientific daters are tackling.

Other questions concern the development of human beings. Were the Neanderthals and Cro Magnons, two groups of early people, once neighbors in the Middle East? When did the first people arrive in North America? When was corn first cultivated? Is the piece of cloth known as the Shroud of Turin the burial wrap of Jesus Christ or is it a fake? What about papyrus (a kind of paper) scrolls found near the Dead Sea, paintings by sixteenth-century artists, and architectural drawings for ancient Egyptian pyramids? How do we tell priceless originals from clever reproductions?

Today, scientists in many fields call upon dating laboratories to help them shed light on everything from the destruction of past civilizations to the likelihood of future global warming trends. At the same time, current dating methods are still being improved. Scientists realize that the ages now assigned to many of the objects in modern museums will become more accurate as better dating techniques are developed.

Each time we successfully establish the age of another treasure from the past, we add to the heritage that all of us share. We also learn a little more about where we come from and where we're going.

2

THE ABCs OF THE
DATING GAME

All scientific dating processes are either direct or indirect. We can directly date items that contain something we know *how* to date. For instance, many plants and animals that lived during the last fifty thousand years can be dated with the radiocarbon method, which you will read about in Chapter Five. This is possible because all organisms (living things) contain the element carbon, which we know how to date.

However, we often want to know how old something is that contains nothing datable. Extremely ancient bones present us with this problem. In such cases, an indirect method is used. For example, we may date a rock or other item found near the undatable bone.

But no scientific dating method is perfect yet. For that reason, investigators generally use two or more methods on a sample. Wherever possible, they also date many samples from the same site. Scientific dating also calls for a great deal of logic and common sense.

THE EASIEST DATES TO FIND ▪ The simplest objects to date are those from the recent past. Written documents are a great help, and they are

available for the past five thousand years. If we find a record written on
[parch...] [Egyp]tian tomb, we have a tool with which to date

date accurately. Take a look at a penny.
[U]nited States, it is marked with the year in
[an]cient coins were dated, too. For instance,
[co]ins were engraved with a certain year in a
find genuine Roman coins, they may help
[wit]h them. All we need to know is how the
[t]o our own.

[...]s good time markers, too. For instance,
in Central America, inscribed dates on
A.D 250 to 900. By comparing our own
[calend]ar, we can tell in what year certain mon-

LET'S GET SERIATION • Another valuable technique for dating objects from the recent past relies on cultural changes. This technique is called seriation (from the word *serial*), and we all see plenty of it in our daily lives. Seriation is based on the idea that people are always changing.

Our current styles in most things—from electronic equipment and airplanes to clothes and tools—keep on changing. We can arrange the products in use today in a time sequence based on the changes in their styles. This kind of sequencing is called a chronology. Using the same principle, we can build past chronologies based on the order in which people made ancient items.

Pottery is a good source for past chronologies, since it can easily be made in many different styles. It also survives burial in soil for many centuries. Archaeologists (scientists who study ancient human cultures) work out chronologies from the order in which ancient people made pottery of certain shapes, designs, or colors. Later, if similar pieces are

Airplanes can provide a modern example of *seriation*. These magnificent flying machines have come a long way from the *Flyer*, flown at about 30 miles per hour in 1903, to the Lockheed SK-71 A jet, which set the air speed record of 2,193.167 miles per hour in 1975. If you saw a museum exhibit on "The History of Airplanes," you could follow their development by noting stylistic changes in planes. You would also find some changes that are limited to particular countries or airplane manufacturers. Although your models might not have year dates, you could still develop an eye for which planes were made before, after, or during the same time period as others.

*Different styles in American Indian pottery help us
to date when each kind was made relative to other kinds.*

found, they are assumed to be as old as the pieces they resemble. This method won't yield an age in years. However, it can tell us if something was made before, after, or at the same time as something else. In other words, it gives a relative date.

But relative dating is harder than it seems. Experts must decide, for instance, if two styles vary because the items were made at different times or in different places. As an example, two water jugs that were crafted during the same time period may look totally different, because their potters came from different locations and cultures.

Then, too, scientists need to decide whether an object was used at the time it was made. It might have been found by or handed down to later generations. But these questions do not discourage the experts. Rather, they see each dating assignment as a challenge, a detective story waiting to be solved.

EARLY DATING OF
DINOSAURS

Now that we've explored the general principles for dating the recent past, let's move further back in time and tackle more challenging finds, such as dinosaur bones.

Dinosaurs were a group of reptiles that came in many shapes and sizes. Some, such as *Tyrannosaurus rex*, were huge. Others, such as *Compsognathus*, were no larger than a hen. Some dinosaurs were plant eaters, others ate meat. Some had special features, such as beaks or plates; others had horns or fins. Over the past 170 years, researchers have described about 500 different kinds of dinosaurs. Many more dinosaurs still remain to be dug up. These diverse creatures dominated animal life on land for some 140 million years, yet not all kinds of dinosaurs lived at the same time.

Today, we think we know which dinosaurs lived when, before they died out about 65 million years ago. But how did we gain this knowledge? After all, contrary to what some cartoons show, there were no people around during the age of dinosaurs to record their existence. Instead, most of what we know about them comes from the clues dinosaurs left behind—their fossils (from a Latin word meaning "dug up").

Dinosaurs came in a huge variety of sizes and shapes. Shown at left is *Struthiomimus*, from the Mesozoic era.

Dinosaur fossils (top and left) are put together to form an approximate reconstruction of what the animal's skeleton looked like. Tracks, or footprints (right), of extinct animals are also useful to paleontologists.

Fossils are the hardened remains of dead plants and animals. They can also be the footprints and tracks of animals. Most fossils are sea creatures washed downstream, often in floods, along with rocky material that buries them when the rushing water sweeps past. Millions of years later and under special conditions, some are found in their natural state and others are hardened into stone. Dead land animals and plants can also get swept along by raging floodwaters and be buried. This makes land animals that dwell near water, such as dinosaurs, good fossil candidates.

We can only imagine what the skin of the dinosaurs looked like. Skin and flesh decompose quickly after an animal dies. But in some places, the less fleshy parts of the dinosaurs were preserved as fossils.

Paleontologists are scientists who study fossils. In 1842 an English paleontologist, Richard Owen, identified some strange fossils that

What killed the dinosaurs? Nobody knows for sure, but the event took place at the so-called Cretaceous-Tertiary extinction boundary. This name refers to the time when the Cretaceous period ended and the Tertiary period began. At that time the dinosaurs (as well as up to 65 percent of all plants and animals on the earth) vanished. Some scientists think that a meteorite impact caused dramatic climate changes that killed the dinosaurs off suddenly. Others say that the extinctions took place over millions of years and were due to other factors, including gradual climatic changes. Today, scientists study this boundary for clues about what happened 65 million years ago to so change the world.

people had been finding for many centuries as belonging to a unique group of reptiles. He coined the name *Dinosauria,* which in Greek means "monstrous lizard." Although dinosaurs were not lizards, and some were probably rather gentle, the name stuck. Soon fossil hunters in many countries were searching for dinosaur bones.

But at that time nobody could tell how old dinosaurs were. The best they could do was give relative dates. To do that, they used data collected by geologists (scientists who study the earth's physical features).

DATING DUO: ROCK LAYERS AND INDEX FOSSILS ▪ Here are some quick facts about rocks. They should help you begin to see how geologists and paleontologists aid each other in dating their discoveries.

The earth has three main kinds of rock: metamorphic, igneous, and sedimentary. Metamorphic rocks are sedimentary or igneous rocks that have been greatly changed, usually by pressure or high temperature. Igneous rocks are formed from matter that was once hot enough to be liquid. Sometimes the liquid (or molten) rock cools underground. Other igneous rocks come from lava, the liquid rock that shoots or flows out of volcanoes, then hardens as it cools on the earth's surface. Some igneous and metamorphic rocks are vital to twentieth-century dating processes.

However, most fossils are found in sedimentary rock. This type of rock, which makes up two-thirds of the earth's surface, is composed of particles (sediments) in a range of sizes. These include boulders, pebbles, sand, silt, and clay. The particles form as older rocks break down as a result of the natural forces of weathering—mainly erosion due to water, ice, and chemical changes.

Most sedimentary rocks form in shallow areas near the shores of oceans and in lakes after rains wash particles downstream. After millions of years, more rain, and tons of additional matter pressing on the particles, they are squeezed together into a layer of sedimentary rock.

Granite (left) is an example of igneous rock. However, most fossils are found in sedimentary rock, such as sandstone (right). The exposed rock layers of the Grand Canyon (below), carved out by the Colorado River, contain the earth's most complete geological record, with rocks dating from more than 250 million to 3 billion years ago.

Later, still more rains cause new layers (or *strata*, from the Latin word for "layers") of sedimentary rock to form. The largest grains of each layer fall to its bottom. They become boundaries between sections of finer grains.

Sedimentary rock layers can be found in many places, including sites where highways have been cut through hills or where wind and rain have eroded mountains. The Grand Canyon in Arizona is an excellent example of natural layering. It consists of many sedimentary rock layers that were formed millions of years ago when the land was under water. The water eventually dried up. Long afterward, the Colorado River cut through the layers of rock, exposing a living history of the earth's past.

STRATA IN A JAR

Materials: Small glass jar with a tight-fitting cover, fine sand, coarse sand, water.

1. Put ½ cup of fine sand into the jar.
2. Add ½ cup of coarse sand.
3. Fill the jar ¾ of the way with water.
4. Cover the jar and shake it for a minute until the water gets muddy.
5. Leave the jar alone until the water clears, then look closely at the sand.

This gives you an idea of how a layer of sedimentary rock forms, with the larger grains, such as pebbles, on the bottom, and finer grains, such as sand, above.

This record exists because sedimentary rock layers generally form with the oldest layer on the very bottom and the youngest on the top. You might see something like this at home if you collect newspapers in a pile for the recycling center. Your oldest newspapers are probably on the bottom of the pile, while your most recent issues are on the top.

Over the past two centuries, geologists have developed a geologic time scale, based on the order in which sedimentary rock layers formed. This scale divides earth's history into large stretches of time known as eras. The eras, in turn, are divided into shorter time spans, called periods. The last two periods are subdivided into still shorter time spans, called epochs. (Smaller divisions, called formations and based on rock types, are used for local geology but aren't necessary to consider here.)

Although the geologic time scale could not give calendar-year dates to any fossil find, it did allow fossil hunters to put their discoveries in a relative order. And once scientists began using the geologic time scale, they found that they could get relative dates from more than just rocks.

By the early nineteenth century, paleontologists had realized that each rock layer includes its own unique fossils. These are the remains of animals that existed only when that rock layer formed. Scientists call these one-stratum remains index fossils.

Nature is not neat, and major catastrophies, such as landslides and volcanic eruptions, upset the time capsule in the rocks. Some human events, such as mining excavations, also shake up the earth's neat layering of sedimentary rocks. However, geologists have worked out ways to compensate for these disturbances, so that their geologic time scale can remain accurate.

Ancient marine animals, such as this ammonite from the Mesozoic era, serve as index fossils to help scientists date other fossils.

After paleontologists connected index fossils with a certain rock layer, they reasoned that other fossils found near them came from the same time period. Index fossils were valuable to geologists, too. They could determine the relative age of rocks by identifying the index fossils embedded in them.

Today, the use of index fossils is dwindling, since other dating processes offer more accurate results. But in the nineteenth century index fossils were very important. Fossil hunters gave dinosaurs and other ancient creatures a relative place in earth's history based on the rocks and index fossils their bones were found with.

As more and more dinosaur fossils were dug up, they came to be identified with certain geologic periods. For instance, the fossils of *Plateosaurus* were assigned to the Triassic period, those of *Brontosaurus* to the Jurassic period, and those of *Tyrannosaurus rex* to the Cretaceous period. Eventually, finding an unidentified dinosaur's bones near those of a known and dated dinosaur made dating the new find possible.

However, no matter how many old bones nineteenth-century fossil hunters unearthed, none of them could be assigned calendar-year dates. That had to wait until the twentieth century, when "dating clocks" were found in many of earth's rocks!

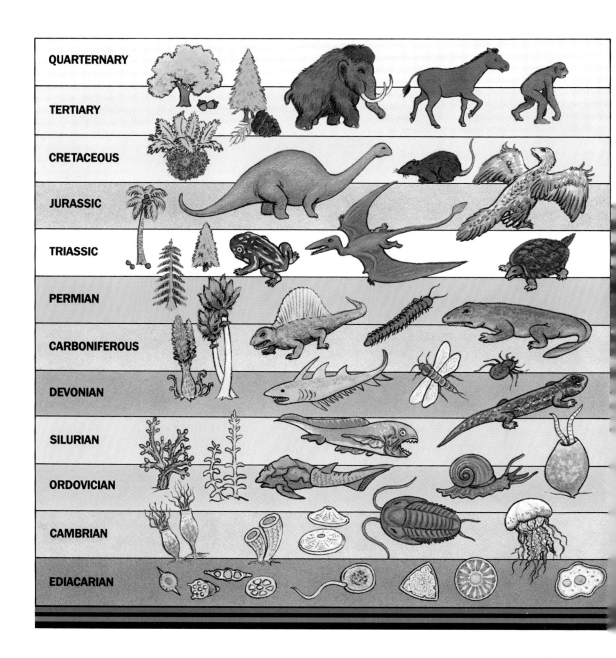

QUARTERNARY

TERTIARY

CRETACEOUS

JURASSIC

TRIASSIC

PERMIAN

CARBONIFEROUS

DEVONIAN

SILURIAN

ORDOVICIAN

CAMBRIAN

EDIACARIAN

GEOLOGIC TIME SCALE WITH CHARACTERISTIC LIFE

Period	Epoch	Characteristic Life	Millions of Years Ago
QUARTERNARY	Holocene	Modern people	.01
	Pleistocene	Mammoths, early people	2
TERTIARY	Pliocene	Apelike creatures	6
	Miocene	Apes, mammals flourish	22.5
	Oligocene	Early apes, modern mammals	38
	Eocene	Early horses, modern plants	55
	Palocene	Mammals multiply	65
CRETACEOUS		Last dinosaurs, small mammals, first flowering plants, new insects	144
JURASSIC		Dinosaurs, first birds and mammals	213
TRIASSIC		First dinosaurs, crocodiles and turtles, forests, many ammonites (sea animals)	248
PERMIAN		Mammal-like reptiles, first trees	286
CARBONIFEROUS		First reptiles, amphibians flourish, large arthropods, early plants (clubmosses, ferns, horsetails)	360
DEVONIAN		First amphibians, sharks, first flying insects	408
SILURIAN		First land plants, fishes with jaws, air-breathing land animals	438
ORDOVICIAN		First fishes, abundant trilobites (sea animals)	505
CAMBRIAN		Sea animals without backbones (invertebrates)	590
EDIACARIAN		Multi-celled living forms increase	670

Most periods were named for the places where their rock formations were first found. Epochs are named for marine fossils that lived during that specific span of time.

4

RADIOACTIVITY AND
MODERN DATING

In the late 1890s, scientists discovered an astonishing fact. Certain ordinary elements are radioactive. In other words, they emit invisible particles and waves. This happens because these elements have extra subatomic particles that make them unstable, so that over time they lose some of their energy. In the process they change into other, usually more stable, elements. Scientists call this process "decay."

The first element found to be radioactive was uranium, which often changes into a form of lead. By the early twentieth century, scientists were experimenting with several radioactive elements. They discovered that half of a radioactive element decays into another element in a specific amount of time, which they call the radioactive element's half-life. In the element's next half-life, half of what remains decays, and so on. The original radioactive element is known as the "parent," and the element it becomes is called the "daughter."

From this, scientists were able to devise ways of using radioactive elements in rocks and minerals as geologic "clocks" to come up with dates in calendar years. These are known as absolute dates, although they are actually approximate dates, not exact dates. A margin of error is always allowed for when using radioactive elements for dating.

Scientists could now calculate the age of a radioactive element as long as they knew the element's decay rate and the ratio of parent element to daughter element. The more daughter element the sample had, the older it was. By 1913, a British geologist, Arthur Holmes, had published a new and more useful geologic time scale. In it, the various geologic periods and epochs were assigned specific ages based on radioactive dating.

ATOMIC THEORY AND RADIOACTIVITY

All things are made up of tiny particles called *atoms*. The atoms can be divided into even tinier *subatomic* particles. Every atom has a central core, called a *nucleus*. Inside the nucleus are *protons* and *neutrons*. Other small particles, called *electrons*, move around the nucleus. The atom has a negatively charged electron for each positively charged proton. The neutron has no charge.

The *atomic weight* of an element equals the sum of its electrons, protons, and neutrons. However, different forms of the same element may have different numbers of neutrons in the nucleus. These are called *isotopes* of the element; isotopes have a different atomic weight than the original element.

Although most atoms are stable, isotopes are often unstable. This is because an isotope's extra neutrons give it extra energy, which it emits as radiation in the form of particles or waves. Elements that emit radiation are said to be "radioactive."

The most useful radioactive (or radiometric) clocks are those that contain uranium 238, which changes into lead 206; rubidium 87, which changes into strontium 87; and potassium 40, which changes into argon 40, a gas. These elements are generally found in igneous rocks. Since fossils are located mainly in sedimentary rocks, scientists generally search for datable igneous rocks mixed in with sedimentary rocks.

For fossils over 5 million years old, potassium-argon dating is the most widely used. Potassium is one of the earth's most abundant elements. It is found in rocks and minerals, as well as in bones. In fact, your body contains radioactive potassium that is constantly changing into argon gas. But we can't directly measure the argon in fossil bones, because the argon leaks away into the air.

THE MAIN RADIOACTIVE ELEMENTS IN "ROCK CLOCKS"

Element	Changes to	Half-Life	Age Range
Rubidium 87	Strontium 87	49 billion years	greater than 100 million years ago
Uranium 238	Lead 206	4.5 billion years	greater than 100 million years ago
Potassium 40	Argon 40	1.25 billion years	greater than 100,000 years ago
Uranium 235	Lead 207	.70 billion years	greater than 100 million years ago
Thorium 232	Lead 208	14 billion years	greater than 200 million years ago

Carbon-14 — ?

Potassium 40 has a half-life of 1.25 billion years. This means that in 1.25 billion years from now, half of the potassium 40 on the earth will be gone, having changed into argon 40. The remaining potassium 40 will be half gone in another 1.25 billion years and so on. Scientists can date potassium-rich rocks or minerals by comparing the number of unchanged potassium 40 atoms to those that have been changed into the element argon.

In order to get reliable potassium-argon dates, scientists need samples with the argon locked inside. These are most often found in the liquid rock spewed out during volcanic eruptions. As soon as the rock hits the air and starts cooling down, it loses any old argon in it. The rock's potassium-argon "clock" resets to zero. When the rock hardens, any new argon produced by the ongoing potassium decay is sealed within it. This helps to date the rock as well as the volcanic eruption that produced it.

Volcanic ash as well as rock can be dated. But scientists need samples that haven't been contaminated. Newly spewed rocks and ashes can be contaminated by landing on material from earlier eruptions. Contamination can also take place if material from later eruptions lands on them.

In the potassium-argon dating laboratory, the scientist crushes a rock sample into mineral powder. The radioactive potassium in the sample is measured. Then the sample, in an airtight container, is heated until it melts, releasing its argon gas. This may amount to no more than a few trillion atoms. (A trillion is 1 followed by 12 zeros.)

The potassium-argon mass spectrometer, first built in the 1950s, measures the ratio of potassium to argon.

Now the scientist uses the basic tool of radiometric dating, a sensitive machine known as a mass spectrometer. If you know the amount and decay rate of your parent element, the mass spectrometer allows you to find your sample's age by measuring how much of the daughter element is in the rock sample. In this case, the older the rock, the more argon it will contain.

Today, by using laser beams, reliable dates can be obtained from much smaller samples of potassium than ever before. In some cases, only a single crystal of a mineral is needed.

After the potassium-argon dating technique was developed, numerical ages could be assigned to dinosaurs as well as to many other ancient animals and plants. Dinosaurs were dated by measuring rocks in strata where they were found, as well as in strata above and below them. Since many dinosaur fossils were dug up near index fossils known as ammonites, these ancient sea animals served as another age check.

Potassium-argon dating is also very valuable to anthropologists (scientists who study the origin and development of human beings). They have used it to assign dates to bones belonging to our distant ancestors.

Anthropologists call the earliest true humans *australopithecines*. In 1959, Dr. Mary Leakey discovered a skull and several hundred bones from one such creature. She nicknamed her discovery "Zing" (for the full scientific name of *Australopithecus Zinjanthropus*). Zing's bones were found in East Africa, buried in volcanic rock. Scientists used the newly developed potassium-argon dating method on the rock, making Zing the first of his kind to be reliably dated. He was found to be about 1.8 million years old.

Mary Leakey, with her husband Dr. Louis Leakey, shows the newly discovered jaw of "Zing," found in East Africa. Zing was the first of our early human ancestors to be dated accurately and is also known as the "Nutcracker Man" because he had very large molars.

Left: Moon rocks brought back by the Apollo *astronauts were dated by several radiometric methods and found to be almost as old as the solar system itself. Right: Radiometric methods reveal that some meteorites are the oldest rocks ever dated, going back as far as 4.6 billion years.*

Much older objects have been dated with radioactivity, including moon rocks brought back by the *Apollo* astronauts. Rocks from the lunar flat plains (known as "mares") as well as from the lunar highlands, or mountainous areas, were dated using several different radiometric methods. Some moon rocks from the highlands have been shown to be around 4.6 billion years old.

Even older rocks have been found right on earth! These are meteorites—large rock fragments that reach the earth from outer space. With radiometric dating, we have found that the oldest meteorites are slightly older than the moon's highland rocks. This makes meteorites the most ancient material found so far in the solar system. We suspect that, like the moon, they were formed at about the same time as the rest of the solar system. (EARTH)

Meanwhile, uranium, lead, and rubidium dating have told us that there are igneous rocks nearly 4 billion years old in Greenland and in the United States, while rocks from Australia analyzed with uranium-lead are apparently 4.2 billion years old. (The earliest life-forms preserved in earth's rocks are bacterialike fossils about 3.5 billion years old.)

Radiometric dating has also been used to help us date the sea floor. With a combination of dating techniques, we know now that the age of the sea floor ranges up to 200 million years old. (It is younger than the rest of the earth because new sea floor is constantly being created.) Scientists are now studying chemical changes within layers of deep-sea sediment to map ancient warm and cold periods.

All of the radioactive elements mentioned so far are primordial. This means they were produced at the same time as the earth and have been decaying since then. But there are other radioactive elements that decay as a result of being bombarded with energy from outside the earth. These elements are called cosmogenic, and one of them is today's leading source of dates for archaeologists.

5

COSMIC TIME MARKER: RADIOCARBON DATING

Although potassium-argon remains the major dating clock for dinosaur bones, another radioactive timekeeper is the most common method for finding the ages of many objects that are less than fifty thousand years old. This is radiocarbon dating.

Radiocarbon dating is generally not used on rocks and minerals. Rather, it is used on objects that were once alive or contain something that once lived. These include bone, skin, hair, seeds, grains, and grass, as well as paper, textiles, and peat. Under certain circumstances, shells, tusks, teeth, plaster, wood, and charcoal can also be radiocarbon dated.

Here's how radiocarbon dating works: Almost 99 percent of all carbon is carbon 12, which is stable. Nearly all the rest is stable carbon 13. However, unstable radioactive carbon 14 (or radiocarbon) forms in small amounts when cosmic rays cause it to be produced in the upper atmosphere. The cosmic rays come from the sun and other sources in our galaxy.

Carbon 14 combines with oxygen in the air to form carbon dioxide, which is then absorbed by plants. Animals get their carbon 14 by eating plants or plant-eating animals. During the life of a plant or an animal, carbon 14 decays but is generally replaced at a rate equal to its decay.

This replacement process ends when an organism dies. At that time the carbon 14 trapped within decays until it is all gone. (Carbon 14 has a half-life of 5,730 years.) The age of a sample is calculated by measuring the relative amount of radiocarbon it still contains.

Dr. Willard Frank Libby developed radiocarbon dating in the late 1940s. His original apparatus has undergone many improvements since then. Today's most advanced piece of radiocarbon equipment is a machine called a tandem accelerator mass spectrometer (TAMS). First built in the 1970s, the powerful TAMS permits scientists to collect data from samples more than 1,000 times smaller than earlier instruments allowed. Specimens as tiny as a seed or a postage stamp-sized piece of cloth can now be dated. In fact, one of the most publicized uses of the TAMS was to date a small piece of fabric taken from perhaps the most studied cloth in earth's history, called the Shroud of Turin.

R adiocarbon was used to date the bones of animals found in California's famous La Brea tar pits. (*Brea* is Spanish for "tar," the pit having been named in 1769 by Gaspar de Purtola, a Spanish explorer.) Some 200 animals had accidentally gotten stuck in the natural substance over the past ten thousand to twenty thousand years. The bones of extinct mammals such as mammoths and saber-toothed cats have been found here.

THE SHROUD OF TURIN ▪ A shroud is a cloth used to wrap a body for burial. What made the Shroud of Turin so special was that many believed it to be the burial wrapping for Jesus Christ two thousand years ago.

A fourteenth-century French knight proclaimed he had the 14-foot-long shroud. But he died in battle soon after making this claim and never did say where he had obtained the cloth. Since 1694, it has been displayed above the altar in the Royal Chapel of Turin Cathedral in Turin, Italy.

The startling thing about the shroud is that it retains the faint imprint of a man with his hands folded in front of him. How the image was made remains a mystery. Over the centuries, as scientific dating became more accurate, increasing numbers of people wanted to know the true age of the shroud. If it dated back to the first century, that would make it possible to have belonged to Jesus Christ.

Because the cloth was handwoven from linen, which is made from flax (a plant), radiocarbon dating could be used. But not until the TAMS could we get a reliable age from an extremely tiny piece of the fabric. No one wanted to permanently damage the cloth just to date it.

In 1987, the Vatican, which owns the shroud, permitted three radiocarbon laboratories to date it. Two were in Europe and one was at the University of Arizona (UA) in the United States. Each laboratory received a small sample from the shroud and three control samples from other fabrics that had been dated earlier. The control samples were used to test the accuracy of each laboratory. The shroud sample and each control sample were carefully sealed inside separate stainless steel containers.

At the UA's Radiocarbon Laboratory in Tucson, Arizona, scientists cut each sample into four pieces and put them in safes at four separate locations. Then the following steps were taken with each sample:

1 • It was carefully cleaned.
2 • It was burned in a sealed container to change its carbon into carbon dioxide, a gas.
3 • The carbon dioxide gas was converted into solid graphite powder. (Graphite is the soft form of carbon that we are most familiar with as pencil lead.)
4 • The graphite was compressed into tiny pellets.
5 • The graphite pellets were loaded into the TAMS.

Left: Dr. Austin Long, of the University of Arizona, sits at the computer of a sophisticated piece of radiocarbon dating equipment, the tandem accelerator mass spectrometer.
Right: The Shroud of Turin, bearing the imprint of a man some believe was Jesus Christ, was dated with radiocarbon.

6 • The TAMS bombarded the graphite pellets with cesium, an energetic element that produces a reaction that leads to the release of radiocarbon. Ultrasensitive detectors measured the radiocarbon left in the sample as well as its stable carbon content. Then scientists used a computer to convert the ratio between these two elements into the age of the sample.

The whole world awaited the results. Was the Shroud of Turin old enough to have been Jesus Christ's burial cloth?

In October 1988, the finding of the radiocarbon tests made headline news. The three laboratories found the Shroud of Turin to date from A.D. 1260 to 1390. In other words, they concluded that the shroud was about seven hundred years old and had been made about thirteen centuries after Christ's death.

Despite the test results, the shroud remains an object of worship for many people. One scientist has even suggested that the event of the Resurrection, when Christ is believed by Christians to have returned from the dead, could have caused a burst of light that altered the normal decay of carbon. In years to come, perhaps new dating methods will make the Shroud of Turin headline news once again.

RADIOCARBON AND OTHER PUZZLES • Radiocarbon dating has also helped determine the age and possible authenticity of many other objects from the past, including the Dead Sea scrolls, discovered in 1947. These ancient manuscripts, found in caves near the Dead Sea (part of the border between present-day Israel and Jordan), contain almost all of the books of the Old Testament. The scrolls were written on copper, sheepskin, and papyrus. The latter two could be radiocarbon dated. They were found to have been produced during the first century, making their authenticity more certain.

Radiocarbon also played a role in dating the Step Pyramid at Sakkara in Egypt. This famous tomb of King Zoser was built from architectural plans written on papyrus that was dated to about 2870 B.C. using radiocarbon dating.

And with the TAMS, paintings by old masters can be radiocarbon dated. Paint pigments are not used as samples, since too much of the painting would have to be scraped off. Instead, a small swatch of linen canvas is removed from the outer edge of a painting, to date when the flax it was made from was harvested.

Of course, once a painting is radiocarbon dated, art experts must decide if it is true to the artist's style. There is always the possibility that someone obtained old linen and created a forged work of art on it!

Radiocarbon helped prove the authenticity of this sixteenth-century painting of Sir Thomas More, by the German painter Hans Holbein the Younger.

OTHER RADIOCARBON DATING ACHIEVEMENTS ▪ Since radiocarbon dating can now be used to date objects up to fifty thousand years old, it has been applied to many that help us chart prehistory and the history of civilization as it grew.

For instance, the radiocarbon dating of seeds is helping to map the worldwide development of farming. This record now includes the cultivation of corn, one of the four major crops in the world. Until recently ancient corn specimens were considered too small for direct dating. Instead, their ages were determined indirectly by radiocarbon dating of materials found near them.

But the TAMS made the direct dating of corn possible. In 1990, it was loaded with graphite pellets made with samples of corn cobs from Mexico's Tehuacán Valley, where the oldest known cultivated corn has been found. The results indicate that corn first appeared sometime between 1,500 to 2,500 years later than previously believed, or about 3500 B.C. at the earliest.

Radiocarbon dating is also helping us to decide when and where people first inhabited the Northern Hemisphere. So far, the earliest known Americans were the Clovis hunters, named for the town in New Mexico near where their stone spearpoints were first identified in the 1930s.

Clovis hunters' distinctive stone spearpoints have been radiocarbon-dated to a period beginning around 9500 B.C.

Most likely the Clovis people migrated from Siberia to Alaska via a temporary land bridge that covered the Bering Strait during the last ice age. At that time the sea level was much lower than it is today. Certain finds indicate that these hunters crossed the bridge around twelve thousand years ago and within one thousand years had spread throughout North and South America.

But each year new claims are made for sites in both North and South America that predate those of the Clovis. One such site is at the Meadowcroft Rockshelter, near Pittsburgh, Pennsylvania. There, radiocarbon dating of objects such as charcoal, arrowheads, and the remains of cultivated plants (including walnut) have suggested possible habitation by people twelve thousand years ago or more. However, these dates remain controversial. Work on determining who were America's first people and when they arrived continues, with scientists not yet in agreement.

Radiocarbon also helps us date major climate changes. For instance, scientists once thought that the last ice age ended about thirty thousand years ago. However, radiocarbon dating of seeds and other fossils found in deposits left behind by the glaciers has revealed a different story. We now believe that the ice age actually lasted until about ten thousand to eleven thousand years ago. Previous ice ages have been successfully dated by other radiometric methods.

An exciting new area of radiocarbon dating is in the study of the greenhouse effect, or global warming. Carbon dioxide is called a greenhouse gas because it traps heat in the atmosphere, just as glass traps heat in a greenhouse. Some scientists say that the carbon dioxide we are presently pumping into our air, mainly from the combustion of fossil fuels, is making the earth unnaturally warm. Others consider the warming trend to be natural. A team of U.S. scientists is trying to find out which idea is correct, as part of the federally funded Greenland Ice Sheet Project.

In 1977 workers in Siberia found a baby mammoth that had fallen into an icy pit and been frozen before its body could decompose. With carbon taken from a tiny bit of muscle tissue, scientists were able to date this creature at 27,000 years old.

Greenland's polar ice sheet, thousands of feet thick, formed after many annual winter snowstorms. Eventually, the yearly snowfalls became compacted and turned to ice. Now scientists are drilling through the ice sheet for core samples. Air has been trapped in these samples for as long as 250,000 years.

"One of the mysteries during the ice ages," explains Dr. Alexander Wilson at the University of Arizona, "is that the carbon dioxide level was a lot lower then than in the interglacials [periods between ice ages], like now. So carbon must have gone somewhere in the ice age."

Scientists are radiocarbon dating the carbon 14 in the ancient air found in the ice cores. They are also analyzing the other forms of carbon in them. Their hope is to chart how much carbon dioxide existed during warm and cold episodes in earth's history as well as where the gas came from and went to. This will help them determine how natural the greenhouse effect is.

As you may remember, radiocarbon dating can only be used to date objects less than fifty thousand years old. Therefore, scientists assign ages to the ancient ice cores by using mathematical calculations based on current theories of how glacial ice flows over time.

Radiocarbon dating is a valuable tool for gaining facts about the past as well as insights about the present and future. However, it would not have come this far if not for another major dating method. This process relies on the secrets trees tell. It is called tree-ring dating, or dendrochronology (the study of tree-ring time).

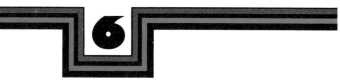

SECRETS TREES
TELL

Have you ever examined a tree stump and noticed the circles on it? These are the growth rings that are laid down annually. By counting the rings of certain trees, scientists can tell many things, including the age of the tree. In fact, tree-ring counting is the most accurate dating tool we have today.

Although people have been curious about growth rings for many centuries, it wasn't until the 1920s that dendrochronology became popular. The "father" of tree-ring science is Dr. Andrew Ellicott Douglass, an astronomer and founder of the University of Arizona's Tree-Ring Laboratory in Tucson.

In the late 1890s, Douglass began to suspect that some kinds of trees could be dated by examining their growth rings. These rings are generally laid down during the tree's annual spring-summer growing seasons. Douglass noticed that certain trees produced growth rings that varied in width from year to year. They would produce a narrow tree ring during a poor growing season and a wide tree ring during a good growing season.

Over the years, trees that lived through the same climatic conditions exhibited a similar pattern of narrow and wide rings. These patterns could "crossdate" with each other. That is, the rings would look very much alike, reflecting the same span of time.

Dr. Douglass set out to build a tree-ring chronology that would go as far back in the past as possible. First, he gathered samples from living trees. Using a hand-powered tool, he extracted a pencil-wide core from the center of each tree. (Core holes are self-healing and do not harm the trees.)

The outermost ring of the living trees represented the current year. By making patterns from the outermost ring inward, Douglass extended his tree-ring chronology as far back as the oldest living sample he could find. Then he dated dead trees, matching each one's last outside ring (which represented the year in which the tree was cut down) with an inside ring of a living tree. He always looked for the one unique place

HOW TO READ A BACKYARD TREE STUMP

Tree stumps are good for sitting on. They are also valuable age indicators. Here's some idea of what scientists do. They count the rings from the outside in. If the tree was just cut, they know that the outermost ring is the current year. Otherwise, they make a graph of the ring pattern and then compare it with a master graph of growth patterns from the same region. If the patterns match at some point, they can be crossdated. If you have some tree stumps near you, see if you can make patterns and compare them.

where the patterns matched. Once he found that place, he could cross-date his samples and extend his tree-ring chronology further and further back in time.

Next, Douglass crossdated wood that came from historic structures in the American Southwest. These included samples from Hopi Indian pueblos, Spanish missions, and pioneer log cabins. After success with these samples, he began testing his technique on prehistoric Indian ruins. Once again, things looked promising.

However, for a long time, Douglass's chronology had gaps, places where he could not match trees with each other. Instead of an absolute chronology for the Southwest, he had what he called a "floating chronology."

Meanwhile, people speculated on the age of the prehistoric Indian ruins in the region. Some said that they were two thousand years old. Others said that they were five thousand years old, or five hundred years old. The only thing known for sure was that the structures must have been built before the Spanish explorers arrived there in the 1600s, because the Spaniards wrote about them.

Douglass began working with archaeologists to date a major excavation site known as the Pueblo Bonito in Chaco Canyon, New Mexico. For ten years Douglass and his team prepared longer and longer chronologies. Yet a gap in the records of prehistoric ruins from the area still remained.

Then one day in 1929, in Show Low, Arizona, two of Douglass's students showed him a new sample they had just found. Douglass labeled the sample HH-39. (HH stood for Haury and Hargrave, the students' last names.) Douglass made graphs of the sample's ring pattern and, after many hours of examination, announced to his startled students, "Gentlemen, we have bridged the gap."

Dr. Bryant Bannister, the former head of the University of Arizona's Tree-Ring Laboratory, calls this "probably the most dramatic moment in the history of American archaeology."

The two edges of a single growth ring are pointed out in this picture of a cross section of a pine tree.

This Douglas fir shows cross dating between two samples of wood from a prehistoric Pueblo Indian ruin in Arizona.

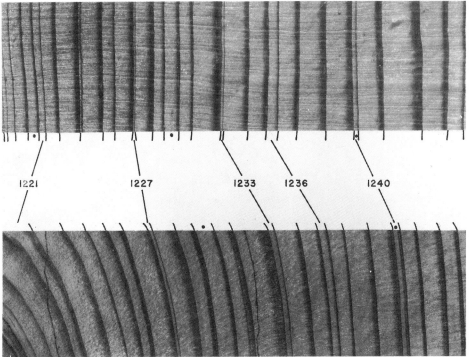

1221 1227 1233 1236 1240

In sample HH-39, Douglass had found the place where the floating chronology could be anchored in time. Now he could extend his tree-ring chronology back through the entire length of the floating chronology. He could show that Pueblo Bonito had been occupied between the eleventh and thirteenth centuries. But he could do more than that. He could pinpoint the calendar ages of prehistoric Indian ruins throughout the Southwest. Since this historic discovery, hundreds of thousands of samples from southwestern prehistoric ruins have been dated with dendrochronology.

Today, tree-ring chronologies in the United States go back nearly ten thousand years. This time range is made possible by a very important and long-living tree: the bristlecone pine. This often bizarre-looking pine is among the earth's oldest living things. By using living and dead samples of it, scientists have built a tree-ring chronology that goes back ninety-six hundred years.

Tree-ring records are used in many archaeological studies. Pieces of charcoal (burned wood), as well as preserved timbers, are dated to tell us many things about the people who used them. Some scientists use these data to determine the behavior of ancient people. For instance, some communities apparently adapted well to sudden climatic changes, while others disappeared from the area. The evidence is a gap in datable wood used for construction sites, indicating a time when no one lived in the area.

But dendrochronology deals with more than archaeological mysteries. It is also used to investigate climate changes on time scales ranging from a few years to thousands of years. In North America, scientists are using tree-ring records to develop a climatic map season by season, going back to the year A.D. 1600. Climatic changes are also being mapped on a global scale with tree rings.

Other changes in the environment can also be studied with tree rings because the chemical composition of each ring can be analyzed.

Pueblo Bonito was the first major ruin in the Southwest to be reliably dated by the tree-ring method.

Some bristlecone pines in the White Mountains of California are 5,000 years old, making them the oldest living things on earth. After scientists determined the age of these trees, the bristlecone pine was put under the protection of the U.S. Forest Service.

As Dr. Bannister explains, "We can look at a ring for 1621 from Arizona, and we can compare the chemical components from 1621 in Siberia, and in Tasmania as well. That makes tree-ring science immensely powerful, both as a dating tool and as a measure of past environmental changes."

This brings us back to radiocarbon, which is absorbed by every living tree. This radiocarbon can be analyzed in each individual tree ring. In the early days of radiocarbon dating, a major error was caught by checking radiocarbon dates with those obtained from tree rings. The error came from a mistaken assumption that the amount of radiocarbon in the atmosphere never changes.

When scientists compared radiocarbon results with tree-ring readings, they were surprised to find that the radiocarbon supply has varied through the ages. Today, radiocarbon dates for the past ten thousand years (as far back as tree-ring dating can go) can be corrected.

But even the highly precise tree-ring research has limitations and requires scientists to be detectives. They must carefully examine every site from which datable wood comes. For instance, a piece of wood may have been reused, making it older than the rest of a particular structure. Or, an old beam could have been replaced with one that was newer than the original structure. Then, too, wooden objects found inside a ruin may be younger or older than the ruin itself. Scientists try to overcome these problems by analyzing several samples from one site and by using more than one dating method whenever possible.

7

SOME DATES ARE
MAGNETIC

The earth is like a giant magnet. It has a positive pole and a negative pole. The space around the earth, as around any magnet, is called a magnetic field.

Today, the positive end of the magnet points to the South Pole and the negative end points to the North Pole. We consider this "normal." But this is not the way it always has been, or always will be.

From time to time, the poles do a flip-flop and reverse, causing the North Pole to become positive and the South Pole to become negative. Scientists call this "abnormal" condition reverse polarity.

Dr. Bernard Brunhes, a French scientist, is credited with recognizing this phenomenon in the early 1900s. To his amazement, he found that when baked bricks cooled, certain particles in them aligned themselves with the earth's magnetic field, so that the brick became magnetized. Brunhes then found that the same thing happened to some particles of lava as they cooled. Even more astounding, some lava flows seemed to be magnetized in a direction opposite to our present "normal" magnetic field.

MAGNETIC FIELDS

Have an adult help you make iron filings from nails (by scraping off an iron nail with a metal file), so that you can see the "lines of force," or magnetic field, around a magnet.

1. Place a sheet of white paper over a bar magnet.
2. Sprinkle your iron filings over the paper.
3. Gently tap the paper. The filings will arrange themselves in lines showing you the direction of the magnetic field. They will also show you that the magnetic force is strongest near the poles of the magnet.

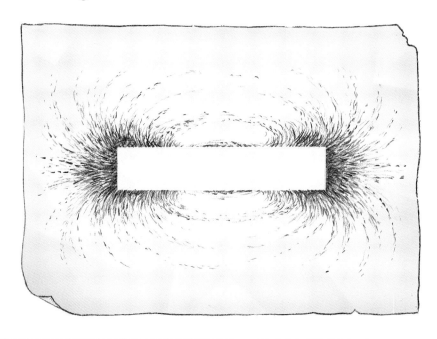

At first, Brunhes's ideas were scoffed at. Other scientists could not imagine the earth's magnetic field changing direction. But gradually Brunhes's theory was confirmed. By the early 1960s it was accepted that the magnetic poles have reversed not just once but many times!

Today, the earth's magnetic flip-flops provide us with a valuable dating tool, since rocks with iron or other magnetic particles recorded the direction of the poles at the time the rocks were formed. Scientists have used these telltale rocks to build a master record of magnetic reversals going back many millions of years. This is known as the paleomagnetic (for "ancient magnetism") calendar and includes reversals that took place at varying intervals, ranging from tens of thousands to millions of years. These reversals can only be put in a relative chronology. However, scientists have assigned year ages to many reversals by radiometrically dating rocks that have magnetic particles.

Today, by using sensitive instruments called magnetometers, we can pinpoint the magnetic period in which certain rocks were formed.

WHAT DOES A MAGNET ATTRACT?

Not all objects are attracted by magnets. You can see this for yourself by holding a variety of objects to a bar magnet. Hold up a paper clip, a pencil, an iron nail, and a penny, and see what happens. You will find magnets attract objects made of iron or containing iron (such as steel, a metal made by adding carbon and other substances to iron). See where the attraction to the magnet is the strongest for the magnetic materials.

This makes paleomagnetic dating useful in many scientific fields, including anthropology.

In the 1970s, scientists in Hadar, Ethiopia, dug up fossils of the earliest known creature that was more human than ape. Nicknamed "Lucy" by scientists, this female African specimen was a major discovery, since 40 percent of her skeleton was able to be recovered. As you may imagine, scientists wanted to be sure that Lucy was dated properly! They used several dating methods, including paleomagnetism.

When scientists analyzed magnetic particles from Hadar's volcanic rocks, they realized that the rocks had been formed during a period of reverse polarity. One such period had occurred between 3.1 and 3 million years ago. An older period occurred between 3.6 and 3.4 million years ago. This helped researchers narrow down Lucy's possible age, which additional dating techniques placed at 3 million years.

In the field of geology, an even more dramatic find came when paleomagnetism was used to test volcanic rocks from the sea floor. Scientists were surprised to discover evidence of reversals along the sides of a central crack in the floor. The pattern on one side of the crack is a mirror image of that on the other side. This indicates that new molten matter comes up from beneath the ocean floor and solidifies, recording the direction of the earth's magnetic field at the time. We call this "sea-floor spreading." Finding out this fact has helped to confirm that the earth's crustal plates move, causing continents and oceans to slowly change position over millions of years.

When magnetic dating is used on archaeological finds it is called archaeomagnetism. Archaeomagnetism generally dates much smaller and more local changes in magnetic direction and intensity than paleomagnetism. It is especially useful in dating ancient firepits and burnt walls that are made of clays containing magnetic particles.

Other materials for archaeomagnetic dating are sediments that contain iron-rich minerals. These sediments are laid down in water. Un-

*Dr. Robert Butler of the University of Arizona loads
an archaeomagnetic specimen into a magnetometer.*

der the right conditions, the minerals in them align themselves, like compass needles, with the earth's current magnetic field.

Scientists are hopeful that this process will become an important dating tool for archaeologists. There are still many areas where there are magnetic-rich particles but no objects for radiocarbon dating nor trees nor wood products for tree-ring dating.

William Deaver of the University of Arizona and other scientists in the southwestern United States are now building a master chronology for that region. They are using tree-ring and radiocarbon dates to stretch their chronology back in time, then crossdating their finds.

Archaeomagnetism is already helping us better understand the Hohokam, ancestors of the southwestern Pima and Papago Indians. In one case, a prehistoric Hohokam site excavated in Arizona was found to be divided into two areas, one for living and the other for holding ceremonies. Dating the clay walls helped to prove that the ceremonial part was built after the living quarters had been abandoned. People had suspected this, but archaeomagnetic dating provided the direct evidence.

Elsewhere around the globe, magnetic research is helping scientists pinpoint many past events, including when pottery was first produced in Japan and China. It is also enabling us to date the ancient movements of earth's crustal plates, on which our continents and oceans ride, as well as prehistoric natural disasters, such as earthquakes and volcanoes. Gradually, as we become more adept at recording precise magnetic changes, magnetic dating may play an even greater role in helping to reconstruct our past.

NEW WAYS SCIENTISTS
GET DATES

Although the dating processes already mentioned are the most widely used today, other processes are also being developed. None is without controversy. Here are some of these dating methods.

HEAT DATING • A release of energy is responsible for the dating process known as thermoluminescence (TL). TL refers to the explosion of energy, in the form of light, when certain minerals that had previously been exposed to radiation are reheated to very high temperatures. These minerals include quartz, calcite, and feldspar.

Scientists can use this method to date flint (a fine-grained quartz used by early humans to make tools), as well as campfire stones from ancient fireplaces. In theory, the heat in a fireplace makes electrons in a sample mineral move toward the center of its atoms, releasing energy as light. Then, after the mineral cools, the electrons slowly reverse direction and move outward from the center. When a TL researcher reheats the sample, the amount of light released—which is based on how long the electrons had been moving outward—reveals the sample's age.

DATING OLD BONES

Relative dating of some fossil bones is possible by comparing certain chemicals in a sample with the same chemicals in other samples from the same site. This is possible because all bones buried at the same time in the same place under the same conditions absorb fluorine and uranium at the same rate from the groundwater where they are deposited. Fossil bones also lose nitrogen at the same rate.

Although this process has many limitations, it made history when it helped to expose one of the greatest scientific frauds of all time: the Piltdown hoax.

In 1911, an amateur archaeologist, Charles Dawson, began finding parts of a skull in the Piltdown gravel pit in England. When the skull was assembled, it seemed to have a modern human cranium (skull section enclosing the brain), with an apelike lower jaw. For forty years people puzzled over this strange find. Some experts called it a fake, but others thought Dawson's "dawn man" was a stage in the development of human beings.

The argument was resolved in the early 1950s. At that time, Dr. Kenneth P. Oakley tested the skull's fluorine and nitrogen content. His tests led him to conclude that the human-looking cranium was much older than the ape-looking jaw. Later tests of the skull's uranium content supported his findings.

Further investigation revealed that someone had doctored a human cranium and the jaw of an orangutan to make them look like ancient fossils. To this day, the culprit behind the Piltdown hoax remains unknown.

This method may tell us a great deal about how Cro Magnons, the first modern humans, fit into the history of human life. For decades, scientists assumed that the Cro Magnon came from what is now Europe or Asia between thirty thousand and thirty-two thousand years ago, replacing the more primitive Neanderthals over a few thousand years. However, new evidence may upset this theory.

Fossils of early modern humans were recently found in certain caves in Israel. But these specimens were too old to be dated with radiocarbon. Instead, scientists used the TL method to date flint samples they found at the site. The flint may have been chipped from tools these early modern humans made as they sat around their campfires. The samples were gradually heated, and the amount of light they emitted was measured.

If the TL results are accurate, they indicate that early modern humans lived in Israel some 92,000 years ago. Furthermore, fossils of Neanderthals found in and around Israel have recently been dated as much younger than the early modern human fossils found nearby. The evidence suggests that Neanderthals and Cro Magnons may have been neighbors at one time in the Middle East. In fact, the Cro Magnon may have originated in this region, then journeyed north.

Although scientists may continue debating how to interpret these dating results for many generations, one thing is clear. Each new find we can date challenges our theories and keeps our minds open to new possibilities.

FISSION TRACK DATING ▪ The release of energy is the basis for yet another dating method. This one relies on crystals known as zircons. People are most familiar with large zircons, which are semiprecious stones used in jewelry. But zircons come in many sizes. Those found in volcanic ash are tiny in size. Yet these tiny zircons can be dated by the scratchlike marks within them. These marks are lines left by "explosions" of uranium.

As mentioned earlier, uranium 238 is a radioactive element that decays into lead at a slow but steady rate. However, a very small number of uranium 238 atoms undergo a different kind of decay, called fission. During fission, the uranium atoms explode, one by one. A tiny energy "pop" accompanies each explosion and leaves a faint line on the zircon. This line is called a "fission track."

When dating a zircon, scientists first clean and polish it to remove any scratches on the surface. (They don't want to mistake a blemish for

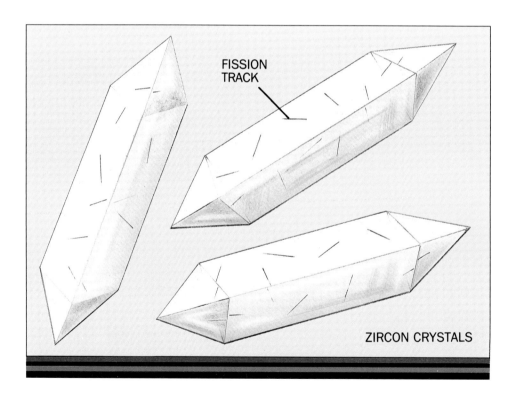

FISSION TRACK

ZIRCON CRYSTALS

a fission track.) Then the zircon is treated with a chemical to make its fission tracks visible under a microscope. These tracks can be counted to reveal the zircon's age.

Although fission-track dating is less accurate than potassium-argon, it serves as a useful substitute when potassium-argon can't be used. It also helps to confirm dates obtained from potassium-argon dating.

OBSIDIAN, THE DATING GLASS ▪ Obsidian is a dark, shiny volcanic glass from which prehistoric tools and weapons were made. Over time, and with exposure to moisture, water reacts with the obsidian to produce a distinctive layer on the surface of the glass. Eventually, this layer sinks deeper and deeper into the glass.

Obsidian's water layer can be examined with a microscope. Measuring the thickness of this layer helps to determine how old the sample is. One place where obsidian dating proved helpful was on Easter Island in the Pacific Ocean, where giant stone statues with huge heads line the coast. Chips of obsidian used in carving the statues have been studied to figure out the time period over which these mysterious figures—thousands of years old—were made.

Although obsidian dating is widely used, it has many limitations. Several slices from the same specimen must usually be measured in order to get accurate readings.

VARVES ▪ Much scientific dating has also been done by measuring various deposits left behind by glaciers. One such method relies on the rate at which layers of sediment settle in lakes near the edges of glaciers. Changing weather conditions produce unique layers, which are known as varves. Some of these layers form annually and thus can be dated. In Sweden, scientists have plotted a varve chronology to date the retreat of glaciers from ten to fifteen thousand years ago.

THE "CLOCKS" IN OUR GENES ▪ One of today's most controversial but promising dating tools involves our genes, the tiny units in our cells that pass on traits from one generation to another. Since the 1960s, scientists have been using genetic data to learn about our ancestors and what the human family tree looks like.

Genes are primarily made up of a chemical called deoxyribonucleic acid (DNA). DNA contains the instructions for life's processes. Genetic dating is based on the fact that mutations continually take place in DNA. Mutations are random chemical changes that, in some cases, can be passed on to future generations.

Scientists who study genes are called geneticists. Some geneticists assume that mutations accumulate at a steady rate. Therefore, when two closely related populations of living things on a family tree separate into two different lines, the longer the two populations are separated, the more their genes will differ.

In 1967, Dr. Allan Wilson from the University of California at Berkeley and his colleagues indirectly analyzed the mutations in DNA that are responsible for producing a certain protein in the blood of chimpanzees and humans. At that time, anthropologists thought that humans and chimpanzees, both on the primate family tree, had split apart between 15 and 30 million years ago. This was based on fossil findings up to the 1960s. However, according to Wilson's genetic studies, humans and chimps differed so slightly that they must have split apart between 5 and 7 million years ago. At first, these figures were ignored by anthropologists. But as more fossil evidence turned up, anthropologists came to accept the geneticists' dates.

More recently, Wilson has introduced a new theory that challenges many current ideas about where and when modern human beings, *Homo sapiens sapiens*, originated. He and his colleagues have studied many genes in an attempt to trace our ancestry back to a single female from

Deoxyribonucleic acid (DNA) is a chemical substance that carries instructions for the inherited characteristics of all living things. It is found in all living cells, largely in the nucleus. Each DNA molecule consists of sugar, phosphate, and four chemicals called *bases*. The bases are adenine, cytosine, guanine, and thymine.

In the 1950s, scientists realized that the DNA molecule is shaped like a twisted ladder. Sugar and phosphate make up the sides of the ladder, which are linked together by steps of paired bases. This same twisted ladder structure is found in all living things, from people and penguins to petunias and pears.

However, living things differ because they have different genetic codes. This code is found in the four bases mentioned above, which can be arranged in many different orders. The code is passed on from one generation to the next. A change in the code is called a mutation. Most mutations are not passed on, but some are.

The DNA found in mitochondria, parts of the cell located outside of the nucleus, mutates about ten times faster than DNA in the cell nucleus. Some geneticists are using the accumulated mutations in DNA both in mitochondria and in the cell nucleus as a basis for finding out when certain organisms lived.

whom they say we all descended. They call that female "Eve." One researcher says that "Eve" may have been the ten thousandth great-grandmother of teenagers living today.

The probable time of "Eve's" birthdate was found by examining a kind of DNA found in the parts of the cell called the mitochondria. Cells may contain hundreds of mitochondria. These sausage-shaped structures, located outside of the cell nucleus, produce the energy needed to keep a cell alive. They also contain a special DNA inherited only from the mother and which is subject to mutation.

For the study, the mitochondrial DNA from the placentas of 147 women were examined. A placenta is an organ that develops during pregnancy and is expelled after birth. Some of the women in the study were American, with ancestors from Africa, Europe, the Middle East, and Asia. Other women in the study traced their heritage to New Guinea and Australia.

After each woman gave birth, her placenta was carefully studied for mutations in mitochondrial DNA. The scientists found that the new-born babies came from a family tree containing two branches, one of African-only descent, the other African plus everyone else. The all-African group had the most mutations, suggesting that it was the older branch. Yet, it was thought that all the babies' DNA could ultimately be traced back to one ancestor, "Eve."

To calculate when "Eve" lived, scientists assumed that the mutations took place at a steady rate. They then calculated back from the present to see how much time was needed to account for the current genetic diversity in humans. "Eve" was found to have lived in South Africa about two hundred thousand years ago. It is harder to trace genes back through males, but scientists are exploring that trail, too. As this book goes to press, reports are coming out about a possible common male ancestor, also from Africa.

The age and place of origin of "Eve" pose a challenge to most scientists, who believe that modern humans developed much more recently, from one hundred thousand to thirty thousand years ago. This new idea is also opposed by scientists who claim that *Homo sapiens sapiens* evolved in Asia or Europe, rather than in Africa.

Although the geneticists' findings are considered exciting, the reliability of genetic data as a dating tool remains debatable. Anthropologists, along with other geneticists, challenge the assumed steady rate of mutation, as well as certain techniques used in the genetics laboratory.

Meanwhile, other groups are investigating how changes in nuclear DNA can be used to date family trees for humans and other living things. Here, too, scientists begin with assumptions that are questionable. But with the field of genetic research still expanding, we may soon have new ways of unraveling the past with clocks right inside our own genes.

DATING IN THE
FUTURE

You have now read about the major methods, as well as some of the more controversial ones, used by modern scientists to obtain dates, both relative and absolute. Yet all of these methods have their own special problems and limitations.

For that reason, some archaeologists in Phoenix, Arizona, are doing something that might at first seem a bit peculiar. They are excavating the Pueblo Grande ruins, once inhabited by the Hohokam Indians. Half of their excavations will be restored for public exhibition. Now here's the strange part: they are going to rebury the rest of their find. Their hope is that scientists of the next few centuries will re-excavate the ruins and analyze them with better scientific dating methods than those available today.

But what about the more immediate future? Scientists are constantly working to improve our present dating methods. Tree-ring chronology, for example, is becoming more of a global effort as scientists in many countries work to produce detailed records of temperature change.

The future also promises more cooperation between the various dating disciplines. Dr. Malcolm Hughes, director of the University of Arizona's Tree-Ring Laboratory, foresees projects involving many dating specialists. These include international tree-ring and ice-core scientists, as well as researchers who date coral reefs, deep-sea sediments, and more. One goal is a year-by-year picture of the world's climatic changes for the past two thousand years.

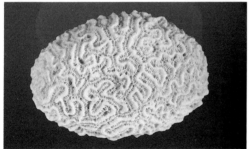

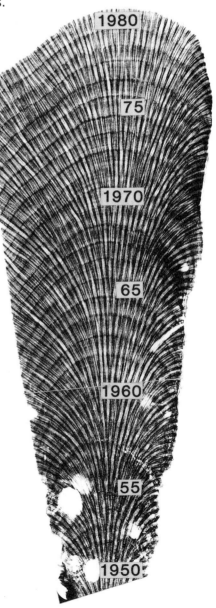

Like tree-ring dating, coral dating relies on signs of growth. Shown above are two common types of coral. This X-ray photo of a skeletal slab of coral (right) shows annual banding, indicating cycles of high (dark) and low (light) density.

— 65 —

What about the development of dating processes that have not yet even been dreamed of? This is a very real possibility. After all, radioactivity, which is the basis for many current dating methods, has been used for dating for only a hundred years. Who knows where tomorrow's dates will come from? We can only guess what secrets await us, hidden inside and under every rock we see.

In the meantime, as the science of dating dinosaurs and other old things grows, scientists are careful to date their samples with more than one method. And they are well aware that many of the time scales and ages we use today will change as we discover more about the past and how to date it. As scientist William Deaver reminds us, "The past is never dead. Not as long as we are alive."

A DATING VOCABULARY

absolute age—the specific age of something, although today's absolute dating methods usually give an approximate, rather than an exact, age.

anthropologist—scientist who studies how people developed physically, socially, and culturally.

archaeologist—scientist who studies material evidence from past human life and culture.

archaeomagnetism—the use of magnetic dating on archaeological finds.

chronology—the arrangement of events in a time sequence that represents the order in which they took place.

cosmic rays—energetic radiation from outer space.

crossdating—matching patterns in the growth rings of trees to determine their ages.

decay—what a radioactive element is said to do when it loses its extra subatomic particles and changes (usually) into a more stable element.

dendrochronology—scientific dating process that relies on counting and crossdating the annual (yearly) growth rings of trees.

deoxyribonucleic acid (DNA)—chemical found in genes that determines hereditary characteristics and is used by scientists to date family trees of living things, including humans.

evolution—the process of gradual change occurring within a species of plants or animals over successive generations.

extinction—the disappearance of a species of plant or animal.

fission track—scientific dating process that involves counting the lines that some radioactive uranium atoms leave on zircon crystals.

fossils—remains or traces of animals or plants that have been preserved in hardened form, usually for ten thousand years or more.

geologic time scale—chart that divides earth's geologic history into segments of time.

geologist—scientist who studies the origin, history, and structure of the earth.

glacier—a river of ice that forms in the earth's polar and mountainous regions.

graphite—a soft form of carbon.

half-life—the time it takes for half of an unstable radioactive element to decay and change into a different element.

ice age—a time when large ice sheets covered Europe and North America. The last ice age ended about ten to eleven thousand years ago. We are now living in what is called an "interglacial period."

igneous rock—rock formed from molten liquid inside the earth or from lava emitted during volcanic eruptions. It hardens as it cools on the surface. Granite and basalt are two forms of igneous rock.

index fossils—fossil animal and plant species that lived for only a short time period and can therefore be used to work out the age of the sedimentary rock layer in which they were found.

isotopes—atoms of the same element with slightly different atomic structures. Some isotopes are natural, while many others are artificially produced.

lava—molten (hot liquid) rock that is generally forced out of cracks in the earth during volcanic action.

magnetic field—the area around every magnet; characterized by having a magnetic force at every point in the region.

magnetometer—instrument used to measure the intensity and direction of the magnetic field when a sample with magnetic particles was formed or burned.

mass spectrometer—instrument used in radiometric dating to measure amounts of parent and daughter elements in a sample.

metamorphic rock—igneous or sedimentary rock that has been changed in structure and composition, primarily by heat, pressure, or both.

mineral—a natural substance in the earth with a characteristic chemical composition.

mitochondrial DNA—deoxyribonucleic acid in mitochondria, certain cell structures that are passed on only through the mother to the offspring.

obsidian—a natural volcanic glass, usually black or banded and useful for a particular dating method.

paleomagnetic calendar—chart that divides earth's history into periods based on the polarity of its magnetic field.

paleomagnetism—scientific dating method that relies on measuring global changes in the earth's magnetic field.

paleontologist—scientist who studies fossils.

radiation—energetic emission from atoms in the form of particles or waves.

radioactive—describes an element in an unstable form, due to having extra particles which it will eventually emit, changing in the process to a different element.

radioactivity—the spontaneous release of radiation from atomic nuclei.

radiocarbon—an unstable form of carbon, also known as carbon 14.

radiocarbon dating—dating process that relies on the decay of radiocarbon to determine the age of objects.

radiometric—describes a device that is capable of detecting and measuring radioactivity.

relative age—age of an object compared to another, not in a year amount but in terms of being older, younger, or the same age as that object.

rock—any naturally occurring solid that makes up the crust of the earth.

sedimentary rock—a kind of rock that slowly forms from clay, sand, silt, or other particles that are squeezed together and under great pressure for long periods of time.

seriation—a relative dating method that arranges something in a time sequence based on the order in which stylistic changes occurred.

tandem accelerator mass spectrometer (TAMS)—the most sophisticated instrument used in the radiocarbon method of scientific dating.

thermoluminescence—scientific dating process based on changes in light emissions caused by the movement of electrons in certain minerals.

varves—layers in certain glacial deposits. Some are laid down annually and can be dated.

FURTHER READING

Asimov, Isaac. *How Did We Find Out About Dinosaurs?* New York: Walker and Company, 1973.

Dixon, Dougal. *Be a Dinosaur Detective.* Minneapolis: Lerner Publications, 1988.

Lafferty, Peter. *Magnets to Generators.* London: Gloucester Press, 1989.

Liptak, Karen. *Pangaea: The Mother Continent.* Tucson: Harbinger House, 1989.

McGowen, Tom. *Radioactivity from the Curies to the Atomic Age.* New York: Franklin Watts, 1986.

Silver, Donald M. *Earth: The Ever-Changing Planet.* New York: Random House, 1989.

Taylor, Paul D. *EyeWitness Books Fossil.* New York: Alfred A. Knopf, 1990.

Vogt, Gregory. *Electricity and Magnetism.* New York: Franklin Watts, 1985.

INDEX

Page numbers in *italics*
refer to illustrations

Bannister, Bryant, 44, 48
Brunhes, Bernard, 49, 51
Butler, Robert, *53*

Carbon, 10, 32-33, 40-41
Carbon dioxide, 32-33, 39-41
Coral growth dating, *65*
Corn, 38

Dating
    coral growth, *65*
    dinosaurs, *8, 9,* 14, *15,* 16-17,
      19-21, 28-29, 66
    fission track, 57-59
    genetic, 60-63
    heat, 55, 57
    humans, 9, 29, *29,* 52, 57, 60-63

Dating (*continued*)
    magnetic, 7, 49-54, *50, 53*
    radioactive (radiometric), 24-31,
      28, 30, 59
    radiocarbon, 10, 32-41, *35, 37,*
      38, 54, 57
    seriation, 12-13
    tree-ring, 7, 41-48, *45, 47,* 54,
      64, *65*
Dawson, Charles, 56
Dead Sea scrolls, 36
Deaver, William, 53, 66
Dendrochronology, 41-48
Deoxyribonucleic acid (DNA), 60-63
Dinosaur National Monument, Utah, *8*
Dinosaurs, *8, 9,* 14, *15,* 16-17, 19-
    21, 28-29, 66
Douglass, Andrew Ellicott, 42-44, 46

Fission track dating, 57-59

Fossils, 14, *15*, 16, *18*, 20-21, *21*, 26, 29, 39, 57

Genetic dating, 60-63
Geologic time scale, 20, *22-23*, 23, 25
Greenhouse effect, 39, 41
Greenland Ice Sheet Project, 39-41

Heat dating, 55, 57
Holmes, Arthur, 25
Hughes, Malcolm, 65
Humans, 9, 29, *29*, 52, 57, 60-63

Igneous rocks, 17, *18*, 26, 31
Index fossils, 20-21, *21*, 29

Leakey, Louis, *29*
Leakey, Mary, 29, *29*
Libby, Willard Frank, 33
Long, Austin, *35*
"Lucy," 52

Magnetic dating, 7, 49-54, *53*
Magnetometers, 51, *53*
Mass spectrometer, 28, *28*
Metamorphic rocks, 17
Meteorites, *30*, 30-31
Moon rocks, 30, *30*, 31

Oakley, Kenneth P., 56

Obsidian, 59
Owen, Richard, 16-17

Paintings, 37, *37*
Piltdown hoax, 56

Radioactive (radiometric) dating, 24-31, *28*, *30*, 59
Radiocarbon dating, 10, 32-41, *35*, *37*, *38*, 54, 57
Rocks, *8*, 17, 19-21, 26, 51

Sedimentary rocks, 17, *18*, 19-20, 26
Seriation, 12-13
Shroud of Turin, 9, 33-36, *35*
Step Pyramid, Sakkara, Egypt, 37

Tandem accelerator mass spectrometer (TAMS), 33-38, *35*
Thermoluminescence (TL), 55, 57
Tree-ring dating, 7, 41-48, *45*, *47*, 54, 64, 65

Varves, 59

Wilson, Alexander, 40
Wilson, Allan, 60

"Zing," 29, *29*
Zircon crystals, 57-59, *58*